Ibrahim Medini

Didactic overview of sheep domestication

Ibrahim Medini

Didactic overview of sheep domestication

Developments in zootechnics andmethods for improving reproductionin sheep

ScienciaScripts

Imprint

Any brand names and product names mentioned in this book are subject to trademark, brand or patent protection and are trademarks or registered trademarks of their respective holders. The use of brand names, product names, common names, trade names, product descriptions etc. even without a particular marking in this work is in no way to be construed to mean that such names may be regarded as unrestricted in respect of trademark and brand protection legislation and could thus be used by anyone.

Cover image: www.ingimage.com

This book is a translation from the original published under ISBN 978-620-6-72714-9.

Publisher:
Sciencia Scripts
is a trademark of
Dodo Books Indian Ocean Ltd. and OmniScriptum S.R.L publishing group

120 High Road, East Finchley, London, N2 9ED, United Kingdom
Str. Armeneasca 28/1, office 1, Chisinau MD-2012, Republic of Moldova, Europe
Printed at: see last page
ISBN: 978-620-3-59542-0

1- HISTORICAL OVERVIEW OF DOMESTICATION (EXAMPLE OF SHEEP)

Helmer (1992) proposed the following definition: "domestication is the control of an animal population by isolation of the herd with loss of panmixy, suppression of natural selection and application of artificial selection based on particular characteristics, either behavioural or structural. Living animals become the property of the human group and are entirely dependent on man".

Time of domestication

The oldest ovine remains have been found in Iraq in strata dated to (-8900 and -8500). Sheep were one of the first species after the dog (-15000 and -12000) and the goat (-9500 and -8500). This period is known as the Neolithic (France : -6,000 to 2,500 BC and Middle East: 6,000 to 4,500 BC), during which time man evolved from a hunter-gatherer to a farmer-breeder.

Place of domestication

The wild species that gave rise to our main domestic species (including sheep) are found in a vast area roughly corresponding to the Middle East today. This vast region, centred on the fertile crescent of ancient Mesopotamia, gave rise to many of the foundations of today's society, including agriculture, urbanism and writing.

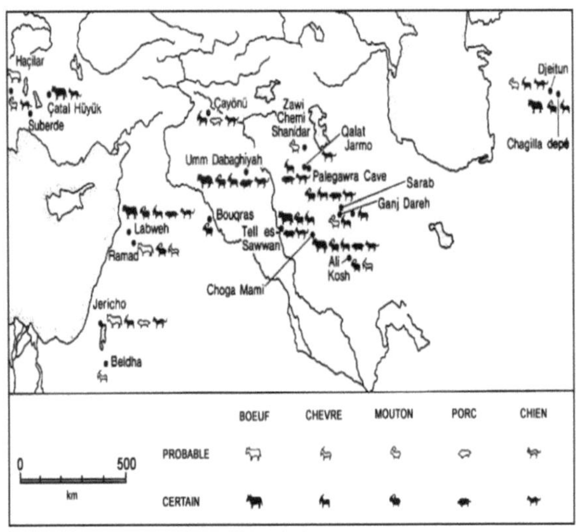

Figure 1: Map of archaeological sites in the Middle East showing the earliest

domestication of animals

2- HISTORICAL OVERVIEW OF ZOOTECHNIE

In an attempt to classify human knowledge, in 1834 the physicist André Marie Ampère subdivided the natural sciences into four main sciences: botany, agriculture, zoology and **zootechnics.**

He defined **zootechnics** as "the science relating to the use or enjoyment we derive from animals, to the work and care by which we procure raw materials from the animal kingdom". He also emphasised the link between the sciences and the technological sciences associated with them:

"Zootechnics is to zoology what agriculture is to botany". In 1843, the term zootechnics was used by Count de Gasparin in the opening lesson of his agriculture course, where he asked the following question:

"Should the education of animals be considered an integral part of agricultural science?

From 1849 onwards, zootechnics became a teaching discipline. Its disciplinary field gradually became structured around reflection on its subject, the production of knowledge and its organisation into a coherent body of knowledge that could be disseminated through teaching.

Animal husbandry has undergone major changes during this century thanks to new knowledge about animal nutrition and genetic progress, leading to the emergence of high-performance meat and dairy breeds such as dairy ewes, which are found in Mediterranean countries.

Sources and players in French zootechnics

Eugène Baudement (1816-1863): "zootechnics is the knowledge of the animal machine". **André Sanson (1826-1902)**: "zootechnics is the science of the production and exploitation of living machines".

Charles Cornevin (1846-1897): "zootechnics is a technological science because it traces the applications that flow from the concepts on which it is based".

Raoul Baron (1852-1908): "Animal science is like the engineer of living machines, whose production and operation he must supervise".

Paul Diffloth (1873-1951): "zootechnics is a new concept of livestock production".

Martial Laplaud (1883-1971): "the aim of zootechnics is to teach the theory and practice of how to make money from domestic animals". **André Max leroy (1892-1978)**: "the aim of zootechnics is to study the rules for making money from animals".

A historical overview of artificial insemination

- The Arabs practised artificial insemination on the mare in the 14th century.

- In 1780, the Italian physiologist Lauro Spanllanzani, professor of natural history in Pavia, injected sperm into the vagina of a bitch which gave birth to 3 puppies after 62 days. - In 1887, the French vet Repiquet inseminated a mare by placing sperm in her vagina using a sponge.

- During the same period (1907), Ivanov published "The Fertilisation of Domestic Animals" (artificial vagina) in Russia.

- In 1944, the first lambs were born through artificial insemination.

- In 1945, the first bovine artificial insemination centre was set up in La Loupe (Eure-et-Loir).

- In 1946, the first calves were born using artificial insemination.

- In 1952, Poldge and Rowson, sperm freezing.

Good reproductive management in sheep farming is reflected in good control of

the components of livestock productivity (figure 13).

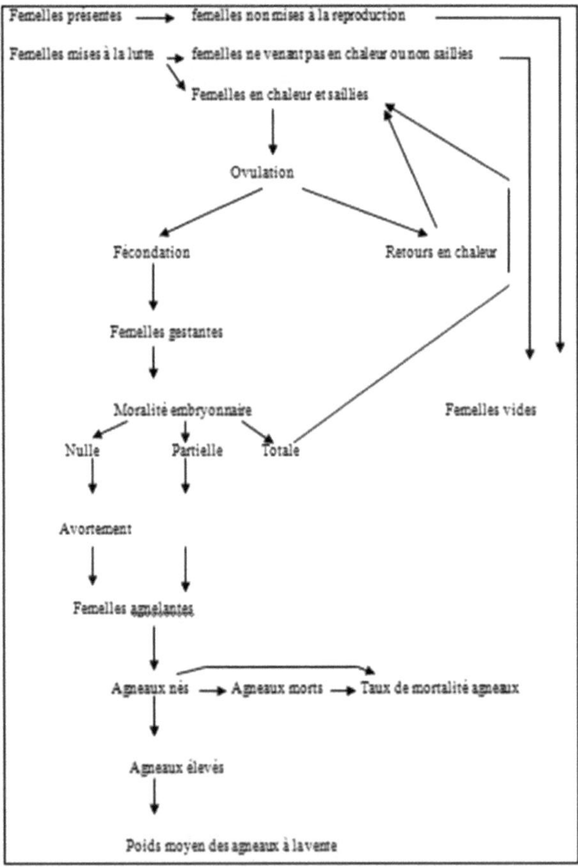

Figure 2: Components of farm productivitysheep (according to OUATTARA

Issif, 2001).

Table 1: Physiological norms in ewes	
HEATS Average duration	: 24 to 48 hours (there are variations depending on breed and age) "adult ewes go into heat for longer than ewe lambs and ewe lambs".
GESTATION (days)	Average length: 146 days (140-152
UTERINE INVOLUTION	It is complete 20 to 30 days after giving birth
OVULATION	This takes place 20 to 30 hours after the onset of heat.
AGE AT PUBERTY female	6 months: when the weight of the corresponds to 40 to 60% of adult weight
CYCLE DURATION	14 to 19 days
GESTATION PERIOD	5 MONTHS ± 1 week
AGE OF MAXIMUM FERTILITY	3 to 6 years
AGE AT REFORMING	5 to 9 years
AGE AT FIRST AGNELLING	10 to 12 months

Table 2: Physiological standards in rams	
AGE AT PUBERTY	6 to 8 months
BREEDING AGE	12 months
FECONDITE	peaks at 2 and a half years of age
FREQUENCY OF USE FOR LA SAILLIE	4 to 5 females per day
AGE AT REFORM	only 5 years to go

- Main types of crossbreeding :

* absorption cross-breeding: successive use of rams of the same breed on dams up to the great granddaughters 7/8.... of the basic flock.

*industrial crossbreeding; the production of this crossbreeding is intended for slaughter. The ram breed is not a requirement.

*the double-stage cross; this is a cross which consists of two crosses between three pure breeds, the first cross of the dams with high lineage rams will provide the prolificacy and precocity characteristics, the cross of the daughters with other rams will provide the conformation and rapid growth characteristics.

4- THE FLUSHING TECHNIQUE

During the struggle period, if the ewes' weight is not sufficient and their body condition score is below 3.5, they are given **'flushing'**. This energy boost during the breeding period helps to improve the flock's prolificacy and average fertility (Jarrige, 1988; Hassoun and Bocquier, 2007). According to Dudouet (2003), requirements during the control period are no different from those during maintenance, but ovulation and the grouping of births are influenced by **Flushing**. The ovulation rate is strongly influenced by feed (figure 3) and therefore by the ewes' nutritional status. (live weight).

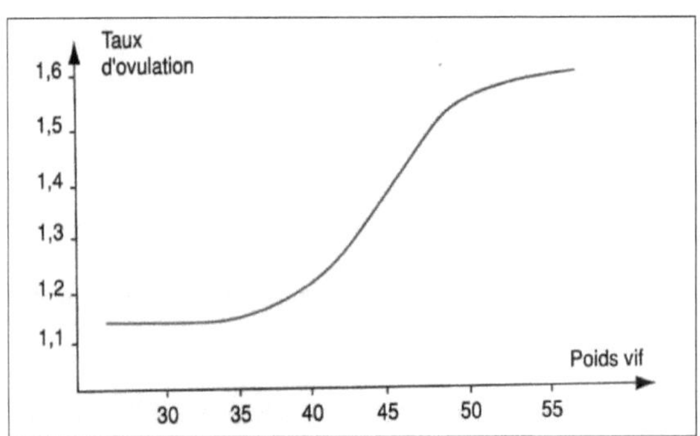

Figure 3: Effect of Flushing on ovulation rate in ewes.

Dudouet (2003) mentioned that there are 2 types of **Flushing** (Figure 4): Pre-oestrus (for 3 weeks before flushing), which improves the number of lambs born by 10-20%, and post-oestrus (for 5 weeks after flushing), which reduces the embryonic mortality rate.

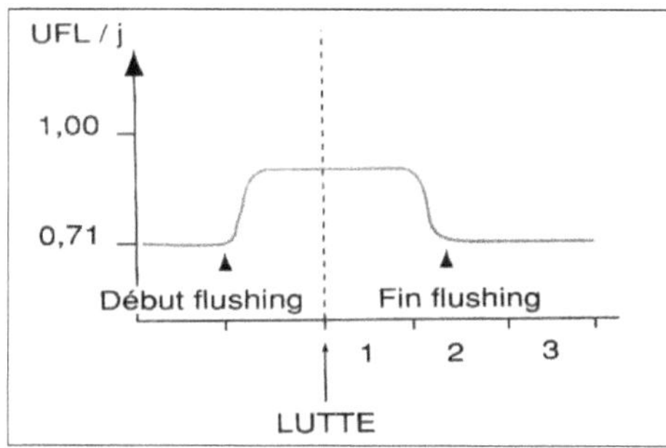

Figure 4: Pre-oestrus flushing and post-oestrus flushing.

According to Fernandez (2003), the **Flushing** technique, which consists of increasing the level of energy intake (supply of 200-400 g per day of barley, for example) three weeks before and three weeks after covering ewes with a body condition index of less than 3, should be used to achieve a good fertility rate.

The same author states that **Flushing** has little influence on the onset of heat, particularly in breeds with deep seasonal anoestrus. Sometimes, a vitamin-mineral supplement helps to improve the effectiveness of **Flushing** during this period, from which **Flushing** improves the level of ovulation and reduces embryonic losses.

Thériez (1984) mentioned that **Flushing** does not increase the fertility rate only in females with average body condition, in fact in lean and fat females **Flushing** has no effect on fertility. **Flushing** reduces the intensity of seasonal anoestrus in ewes by improving their body condition (Khaldi, 2005).

Concentrated overfeeding (**flushing**) increases the live weight of the female at control, which improves her prolificacy (Lindsay, 1995) (figure 5).

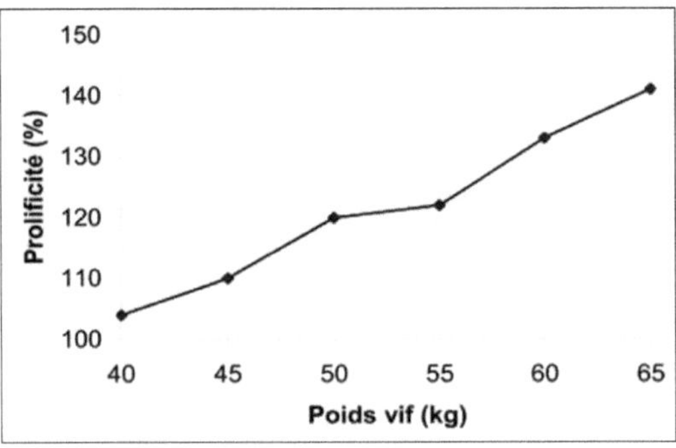

Figure 5: Relationship between live weight at control and prolificacy in Merino

ewes (Linsday, 1995).

Khaldi, (1989), mentioned that **Flushing** can also improve the rate of twin

births.

5- PUBERTY AND FACTORS INFLUENCING THE ONSET OF HEAT IN EWES

Puberty is defined as the age at which the animal becomes capable of producing fertilising gametes, i.e. the first heat in females and the first ejaculation in males. In fact, the age of puberty is 5 to 9 months in sheep, but the appearance of the first heat depends (Dudouet, 2003):

a- Effect of month of birth on onset of first heat in ewes:

Females born at the end of winter can be bred in autumn of the same year, when they are 7 to 8 months old. Late-born females are not bred until they are 12 to 15 months old.

b- Effect of breed on the onset of first heat in ewes :

The age of puberty varies depending on the breed. The Finnish and Romanov breeds can be mated at around 3 to 4 months (table 3).

Table 3: Variation in age at puberty in sheep according to breed (Dudouet, 2003).

Breeds **Age at 1er oestrus Weight at of weight**

	(in j)	Oestrus (kg)	adult
Limousine	250	31	65
Berrichon du cher	261	45	75
Romanov	179	31	65

c-   Effect of temperature on the onset of first heat in ewes:

A temperature of 8 to 9 in the sheepfold brings the onset of heat forward by about a month. **d-Effect of weight on the onset of first heat in ewes:**

Puberty occurs earlier as weight increases (3/4 of adult weight at wrestling).

6- PUBERTY AND FACTORS IN THE VARIATION OF SPERM PRODUCTION IN RAMS

Puberty in males is comparable to that in females. It is a function of the breed, month of birth, food level and environment (light), and even for weight, the male must reach ¾ of the adult live weight at breeding (Dudouet, 2003). He pointed out that sperm production is continuous and proportional to testicular weight. Hence this production is a function of several factors. The sperm production of a young ram is lower than that of an adult ram, due to the age effect. Furthermore, sperm production in rams is at its highest during the short-day period, when the testicles are at their largest. The ram's state of health therefore affects the spermatogenesis process, bearing in mind that the volume of an ordinary ejaculate is on average 0.9 ml, with a sperm concentration of 1.5 to 6 million per millilitre. Puberty occurs earlier the higher the weight (3/4 of the adult weight at wrestling).

7- THE SEXUAL CYCLE OF THE EWE

Sexual activity in ewes is seasonal and is characterised by a succession of oestrous cycles every 17 days on average, which can be broken down into two phases (Cognié et al., 2007).

- The follicular phase lasts 3-4 days and ends with heat and ovulation,

- The luteal phase prepares the uterus for implantation of the embryo. If the ewe has not been fertilised, this luteal phase is interrupted after 13-14 days to make way for a new follicular phase.

The ewe's sexual cycle is the result of several changes: gamete production in the ovary where the ewe becomes aggressive and receptive to the male (heat behaviour), and hormonal secretion of hormones that regulate the cycle. The ewe's sexual cycle is therefore made up of two components (figure 6) (Dudouet, 2003):

-The Oestrian cycle (heat)

-The ovarian cycle (gamete production and ovulation).

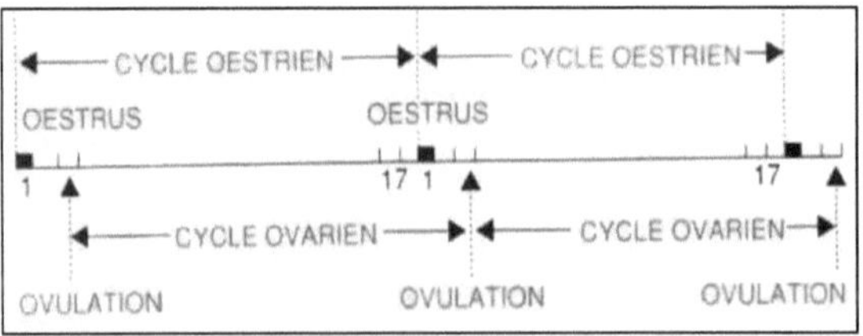

Figure 6: Sexual cycle of the ewe

14

The sexual cycle of female sheep can be divided into three periods.

-Oestrus: this is the period when the female is receptive. The duration of oestrus is proportional to the age of the ewe during this period. She undergoes a number of behavioural changes (agitation, search for the male) and physiological changes (congestion of the vulva, drop in milk production....) (Dudouet, 2003). During this phase, ewes tend to show perceptible signs of heat, such as (Charron, 1986):

- Behavioural changes (increased motor activity, reduced food consumption, tolerance reflex in the presence of a whole male).

- Morphological changes (size, colour) of the external genitalia (increased volume and reddening of the vulva, oedema of the mucosa of the vaginal vestibule, abundant secretion of cervico-vaginal mucosa).

- Reduced viscosity of cervico-vaginal mucus,

- A change in intra-vaginal pH of around 6.5 - A slight decrease in basal body temperature.

Figure 7 below shows the evolution of hormone concentrations during the ewe's sexual cycle.

-The luteal phase: after ovulation, the oocyte is found in the oviduct, and the corpus luteum, formed from the scarred follicle, secretes progesterone to keep the embryo in the uterus, and to block the sexual cycle.

-The pre-ovulatory phase: in the absence of fertilisation, the corpus luteum regresses under the action of other hormones to unblock the sexual cycle.

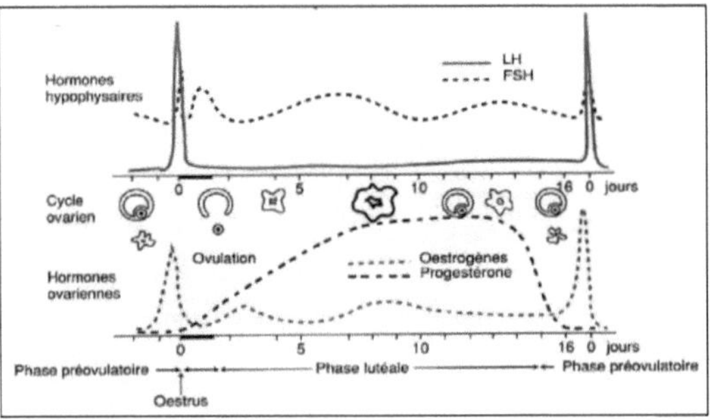

Figure 7: Changes in hormone concentrations during the ewe's sexual cycle.

8- SEASONALITY OF REPRODUCTION IN SHEEP

Reproduction in sheep is characterised by alternating periods of sexual activity and rest (Khaldi, 1984). Sexual activity generally extends from August to February; outside this period, the animals are in anoestrus with a cessation of cyclical ovulatory activity. Dudouet (2003) mentioned that sexual activity in ewes peaks in autumn when day length decreases, which corresponds to the sexual season (Figure 8 a). In the other seasons, when the days are longer, the ewes' sexual activity reaches its minimum or even ceases (figure 8 b), which means that the ewes are in a state of sexual rest, known as seasonal anoestrus.

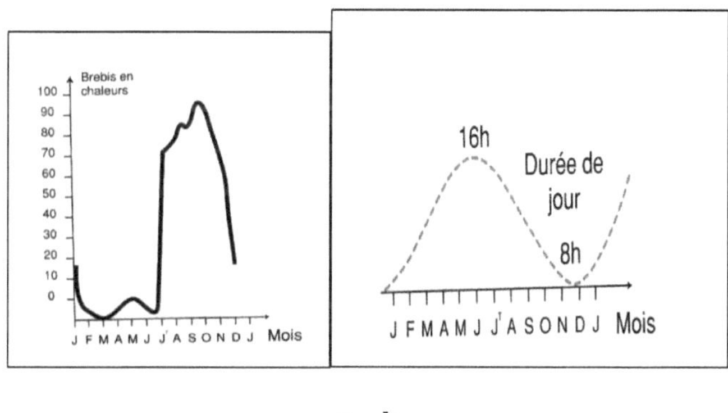

a   b

Figure 8: Seasonal variation in the arrival of ewes in heat for Îles-de-France ewes (Thimonier and Mauléon, 1969).

In rams, sexual activity is a function of the season, with maximum activity in autumn when the days are long and there is 16 hours of daylight (figure 9).

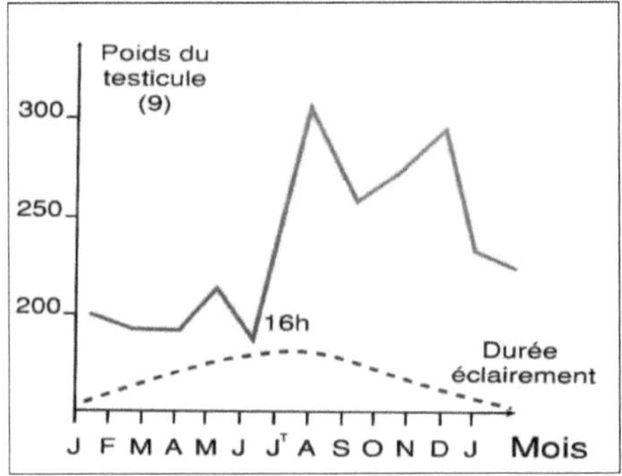

Figure 9: Seasonal variation in the weight of Ile-de-France ram testes (Pelletier, 1971).

According to Zaiem et al (2000), in most sheep populations worldwide, at latitudes above 23°, seasonal variations in sexual activity are dependent on the photoperiod. The short, decreasing days of late summer and autumn stimulate sexual activity, while the long, increasing days of late winter and spring inhibit it.

9- EFFECT OF PHOTOPERIOD ON OVINE SEXUAL ACTIVITY

The length of daylight and its changes are the most influential factors on physiological changes in sheep reproduction. Generally speaking, seasonal reproduction under the effect of this natural phenomenon (short days and long days) results in lambing at an ideal time, with a few exceptions (Photoperiod Reference Guide).

a- Effect of photoperiod on sexual activity in ewes :

Ovine females generally reach their peak sexual activity in autumn, when they are receptive and cyclical, corresponding to the sexual season, or natural breeding period. For the rest of the year, mainly in spring, females are in anoestrus, the period of least sexual activity (counter-season) (Figure 10).

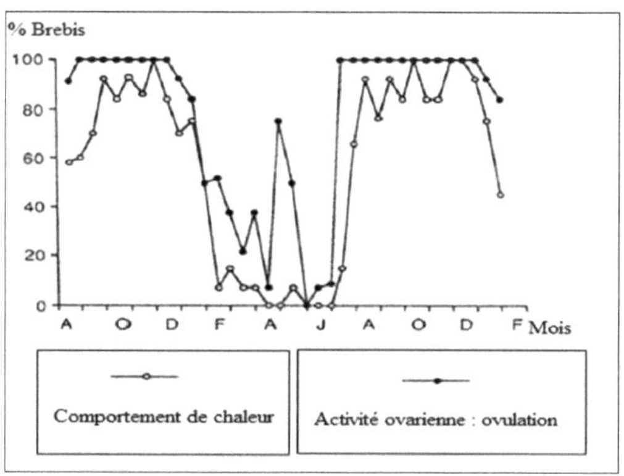

Figure 10: Seasonal variation in sexual activity in Île-de-France ewes.
(Chemineau et al., 1992).

b- Effect of photoperiod on the sexual activity of rams :

Ram reproduction is also seasonal, but the effect of the photoperiod is less intense in rams than in ewes. Sexual activity in males is continuous throughout the year. Endocrine and spermatic sexual activity drops sharply in intensity in spring and summer. This is because the increasing length of the day has an inhibiting effect on the hypothalmic-pituitary-testicular axis.

During the period of short (increasing) days, the concentration of testosterone, FSH and LH falls remarkably. As a result, the concentration of spermatozoa in the semen decreases, with the appearance of gamete abnormalities and a reduction in testicular size and sperm production (figure 11).

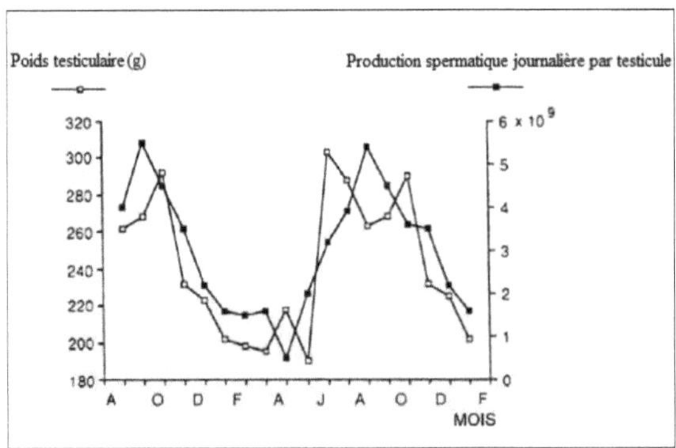

Figure 11: Seasonal variation in testicular weight and sperm production in Île-de-France rams. (Chemineau et al., 1992).

c- **Mechanism of action of photoperiod**

Photoperiod acts on ovine sexual activity via a hormone secreted by the pineal gland, melatonin, which is the hormone responsible for translating the light message.

Animals secrete melatonin during the dark phase (Figure 12).

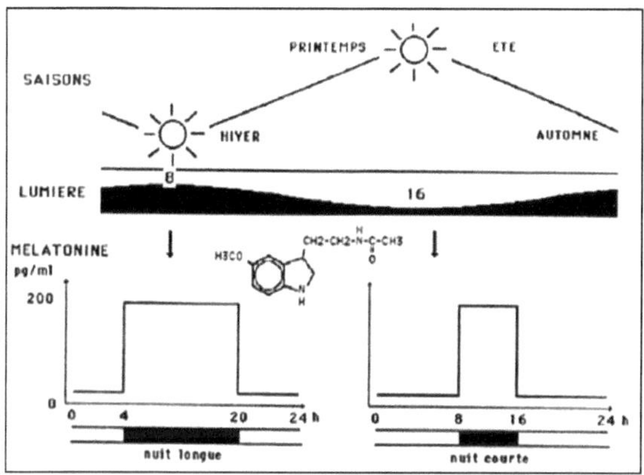

Figure 12: Melatonin secretion pattern as a function of nocturnal duration (Chemineau et al., 1992).

Melatonin is the messenger that enables the nervous system to interpret the photoperiodic signal.The duration of melatonin secretion is directly proportional to the length of the night. Rekik and Mahouachi (1999) have shown that this cyclicity is hormone-dependent through interactions between ovarian hormones (oestrogen and progesterone) and pituitary hormones (LH and FSH). Cyclicity is broken when the female is fertilised, leaving room for the development of the gestation period, which averages 5 months for the species (Figure 13).

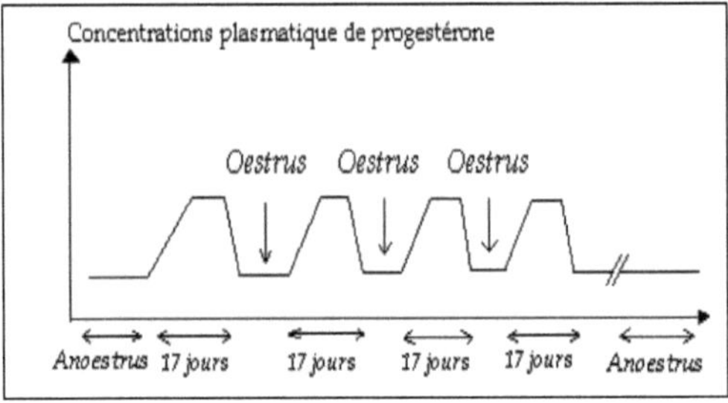

Figure 13: Seasonal variations in the sexual activity of a non-pregnant ewe

(Rekik and Mahouachi, 1999).

Seasonal anoestrus is defined as a period of reduced sexual activity when 50% of females no longer have regular oestrus or cyclical ovulatory activity (Chemineau et al., 2001). For the Barbary breed, seasonal anoestrus is never complete, with sexual activity peaking in autumn and beginning to decline in spring, according to Khaldi (1984). It follows that there are always a certain number of females showing cyclical oestrus behaviour even in spring, a season unfavourable for reproduction, which reflects the shallow anoestrus of this breed (Lassoued and Khaldi, 1995).

<h1 align="center">10- THE RAM EFFECT</h1>

a- Principle of the water hammer effect :

The male effect consists of introducing rams into a flock of anovulatory ewes in spring after a period of absolute separation of at least one month.

The male effect is a natural, economical and effective means of inducing sexual activity in seasonally anoestrous ovine females subjected to off-season fighting (Khaldi, 2007). These females respond to the male effect and ovulate within 2 to 4 days of the introduction of the rams (Khaldi, 1984).

According to Thimonier et al (2000), the male effect is a reliable means of :

- Triggering the ewes' sexual activity during the anoestrus period, particularly in spring, which enables them to produce lambs out of season, with the associated economic benefits.

- Synchronise mating and, to a lesser extent, parturition, making it easier to monitor lambing and create uniform fattening batches.

- To improve the fertility of the flock when the spring grazing period is short in the case of transhumant flocks. (Transhumance is the practice of taking flocks of sheep to graze on pastures).

According to Kennedy (2002), off-season breeding of sheep is increasingly practised as producers adopt accelerated lambing programmes to ensure better year-round market supply.

b- Physiological mechanism of the ram effect :

The introduction of rams into a flock of previously isolated anovulatory females is immediately followed by an increase in the frequency of pulsatile LH discharges. This leads to ovulation. (Thimonier et al., 2000).

Wool and sebaceous gland secretions carry the pheromonal message (Knight and Lynch 1980 cited by Thimonier et al., 2000). The smell of wool can induce an increase in pulsatile LH discharges and ovulation in anovulatory ewes (Signoret 1990 cited by Thimonier et al., 2000).

c- Response to the water hammer effect :

The introduction of rams into a flock of anoestrus (anovulatory) ewes, after a period of separation, induces ovulations not associated with sexual behaviour (heat) in the 2 to 4 days following this introduction, then either normal sexual cycles appear (17 days: case of cyclic females), or a short sexual cycle (6 days: case of non-cyclic females) followed by silent ovulation. Then the normal sexual cycle with heat behaviour sets in, which explains the late onset of heat in females (ewes) whose sexual activity is induced by the introduction of rams, as shown by the two peaks of sexual activity (18 to 20 days) and 24 to 26 days (Khaldi, 1984) (figure 14).

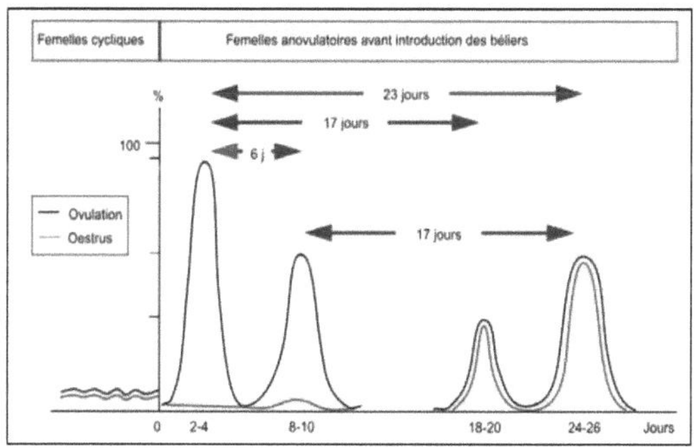

Figure 14: Estrous and ovulatory responses of Barbary ewes to the "male effect". (Thimonier et al., 2000).

According to Thimonier et al (2000), ewe lambs and ram lambs respond less well to the male effect than adult ewes, and the same applies to underfed females. The male effect is all the more effective the closer the end of the sexual season and the longer the interval between drying up and reproduction (Tournadre et al., 2009).It is the intensity of seasonal anoestrus in ovine females that determines their response to the male effect (Khaldi, 1984). According to Khaldi and Lassoued (1991) and Thimonier et al (2000), the proportion of females responding to the male effect varies inversely with the percentage of anovulatory females. This has been demonstrated either by :

- Analysis of the frequency of pulsatile LH discharges in blood samples.

- Analysis of levels of progesterone levels peripheral plasma progesterone levels.

- Direct observation of corpus luteum by endoscopy.

11- THE FIGHT

This is the process that involves fertilisation by bringing rams and ewes together.

a- Natural control :

Natural control can be :

-The rams are permanently with the ewes, with random fertilisation.

- The rams are then removed from the flock and reintroduced on a specific date, hence **the male effect**, which is an effective way of stimulating sexual activity in females.

In both cases, hand-mounted control can be practised, which consists of controlling ewes ewe by ewe and detecting ewes in heat using a vasectomised ram or a ram wearing an apron to prevent mating.

b- Artificial synchronisation :

-Principle

*Inhibition of the triggering of the ewe's sexual cycle by progesterone (P4) found in the sponge, by inhibiting cyclical discharges of pituitary gonadotropic hormones in ewes that are sexually active.

*progesterone prepares ewes in oestrus for the action of PMSG (Pregnant Mare

Serum Gonadotropin).

*The injection of PMSG during the removal of the sponge has several roles such

as ;

*to induce and synchronise heat and ovulation in anostrus females.

*better synchronise oestrus in sexually active ewes

Figure 15: Sponge with progesterone (P4)

Figure 16: Inserting the sponge into the tube

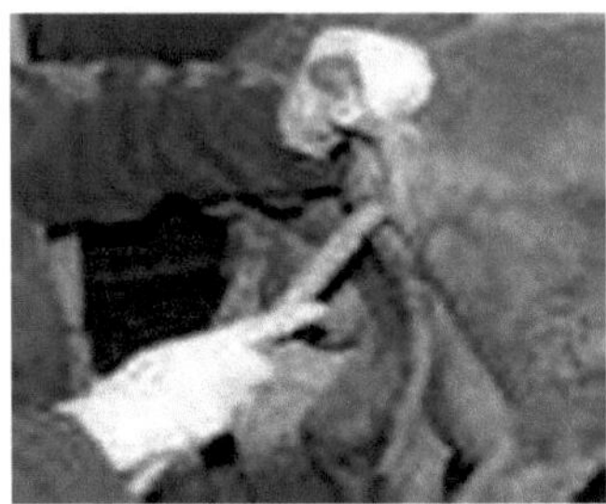

Figure 17: Inserting the tube and using a pusher to push the sponge

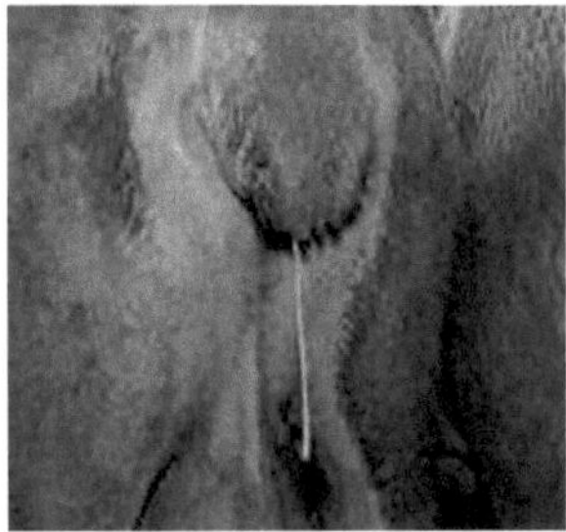

Figure 18: End of operation

c- Fertilisation :

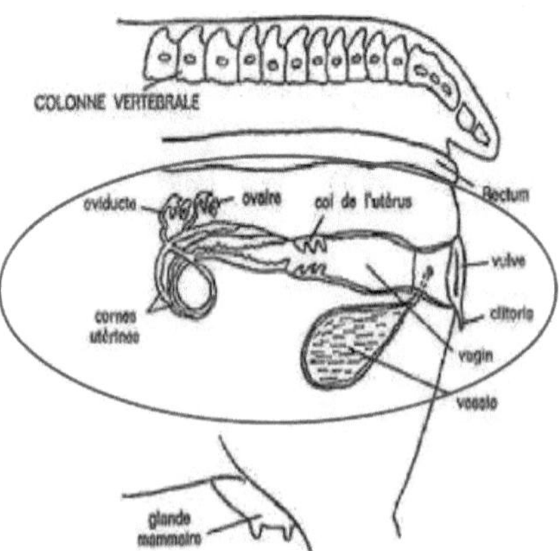

Figure 19: Female genitalia

28

<h1 align="center">12- NATURAL PROTRUSION</h1>

- It can be free without any intervention from the shepherd, in which case the dominant male will mate several times, but the other males will not spread their genetic material, hence the inbreeding phenomenon.

- It can be controlled with the intervention of the shepherd, hence the nomination (monte en main) which consists of presenting the females individually to the males, hence the knowledge of the origin of the offspring.

Figure 20: Free natural protrusion

a- Artificial insemination

Artificial insemination is the technique that enables rapid genetic progress to be made, and is carried out after a series of different operations depending on the physiological stage, such as applying and removing the sponge. Insemination is carried out in a single operation 55 hours after removal of the sponge for adults and 52 hours for ewe lambs.

Table 4: Conditions for AI of French breeds of ewe

Saison Femelles Forme de conservation	Durée de traitement (jours)	PMSG dose (UI)	Nombre de spz ($\times 10^6$)	Volume paillette (ml)	Nombre d'IA	Horaires (heures après retrait de l'éponge)
Anœstrus Brebis Liquide	12	500-700	400	0,25	1	55±1
Agnelles Liquide	12/14	500	400	0,25	1	55±1
Saison Brebis Liquide sexuelle	12/14	400-500	400	0,25	1	55±1
Congelée	12/14	400-500	900	2 x 0,50	1 ou 2	55±1ou 50 et 60
Agnelles Liquide	12/14	400	400	0,25	1	53±1

-Stages of artificial cervical insemination 1- Preparation of the straw

2-Lifting the ewe's hindquarters 3-Slow introduction of the speculum 8 to 10 cm

4-Spotting and introduction of the gun (contains the straw with the semen) -
After about 15 bears of artificial insemination the rams are introduced into the
females for the return to heat for females where fertilisation has failed.

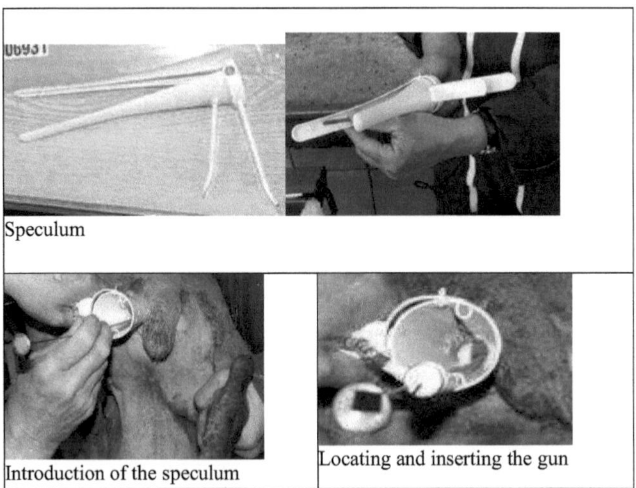

Figure 21: Illustration of the main stages in artificial insemination in ewes -
Intra-uterine insemination :

Intrauterine insemination by endoscopy is mainly used in sheep. It has the dual advantage of increasing fertility when using frozen sperm and using far fewer spermatozoa (around 10 times less than for exocervical insemination). To achieve this, you need to fast for 12 hours beforehand.

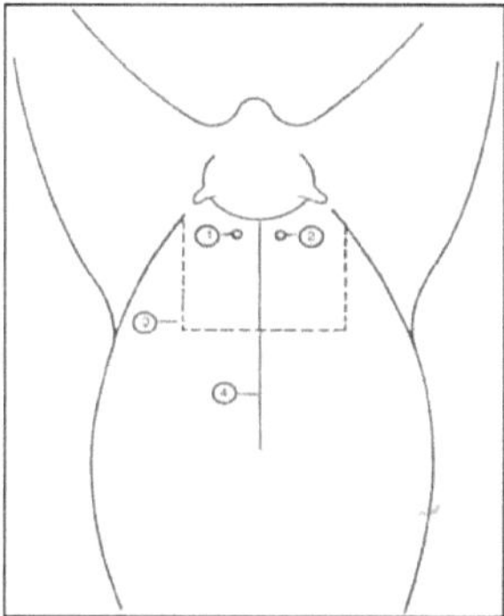

Figure 22: Places where instruments are inserted

Surgical for intrauterine insemination.

1 = Trocar and cannula for optical instruments

2 = Trocar and cannula for artificial insemination instruments

3 = Operating field

4 = Median abdominal line

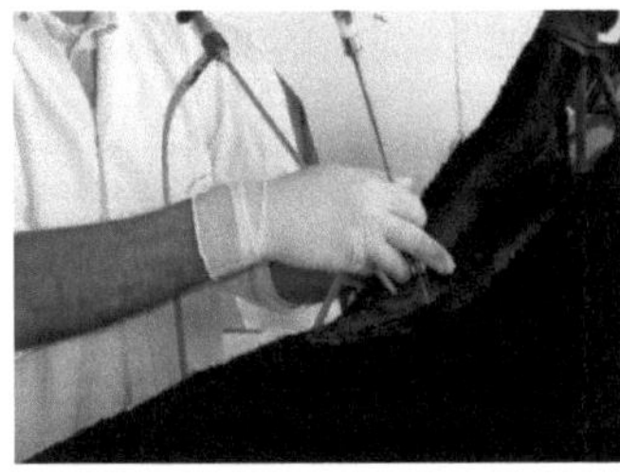

Figure 23: Intrauterine insemination by a Tunisian team on the farm. Ain chalou (Béja north)

b- Gestation:

According to Dudouet (1997), gestation is the time interval between fertilisation and parturition.

Gestation diagnosis is a useful tool for making decisions :

-the reform

-back to the fight

Diagnostic techniques:(Loup Bister)

- use o f a marker carried by the ram (this method is generally unreliable in wet weather).

-observation of the ewes' flanks or udder development.

-palpation of the abdomen

-hormone assays

ultrasound from 40ème days onwards

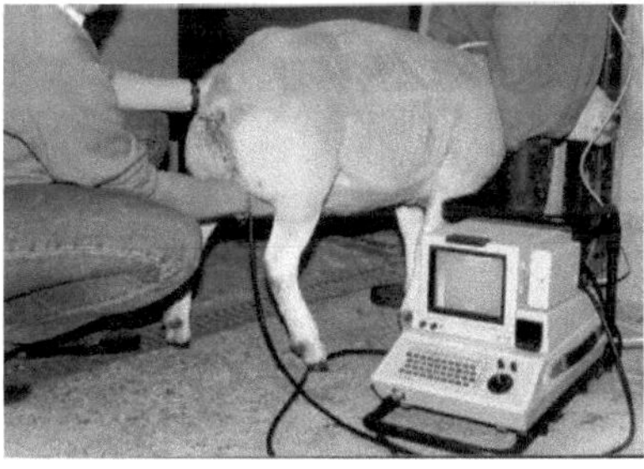

Figure 24: Diagnosis of gestation

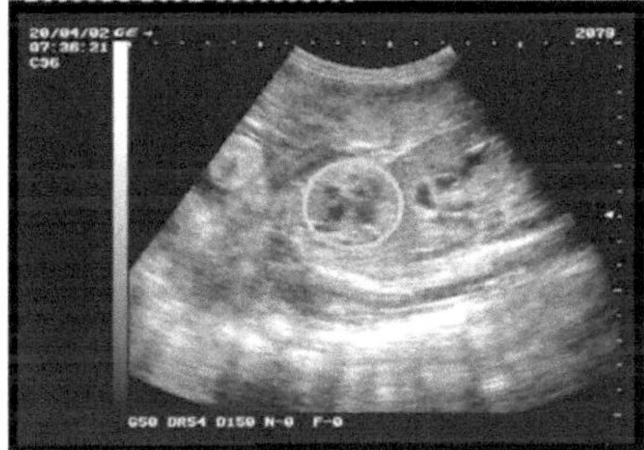

Figure 25: 17-week ultrasound image showing the heart with its 4 chambers (in the circle).

c- **Lambing**

Signs of parturition

-the ewe stands apart from the others.

-The vulva is swollen and the skin folds.

33

-The animal appears agitated and is not eating well.

-a discharge from the vulva occurs a few days before parturition

-the ewe lies down, stretches her neck back to look up and licks her lips

-The ewe pushes to bring out the lamb Lambing takes place in three stages:

-dilation of the cervix

-expulsion of the lamb

-expulsion of the placenta (delivery)

Care of lambs and ewes immediately after lambing

-checking that the lamb is breathing

-disinfection of the lamb's umbilicus

-prevent the ewe from ingesting the hindquarters

- if the lamb is unable to suckle its mother, colostrum should be given via a tube

-The ewe must receive an injection of antibiotic and an egg of antibiotic in the uterus.

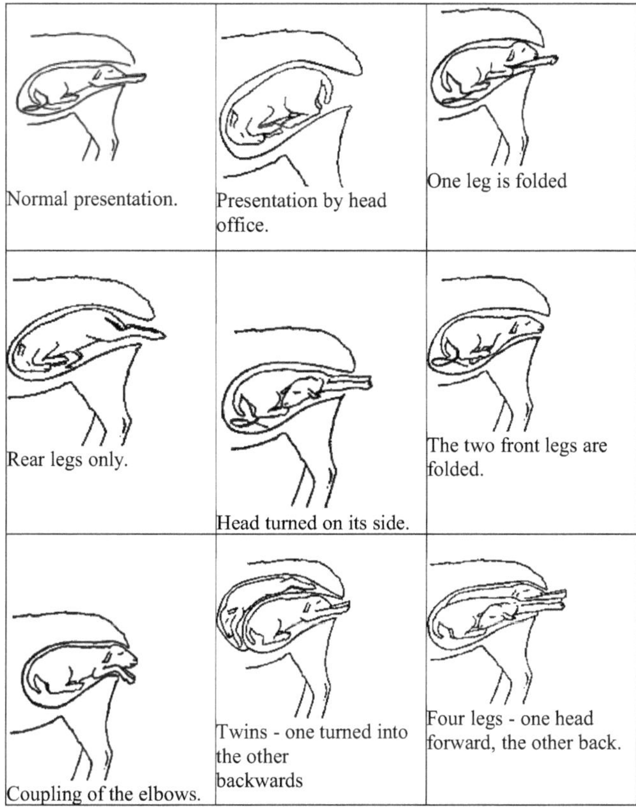

Figure 26: Different presentations of the lamb at lambing time

REFERENCES

Charron G., 1986. Les productions laitières, vol. 1. Lavoisier, Paris. pp 347.

Chemineau P., Daveau A., Maurice F and Delgadillo J A., 1992 Seasonality of oestrus and ovulation is not modified by subjecting female Alpine goats to a tropical photoperiod. Small Rum. Research 8.

Chemineau P., Cognié T., Thimonier J., 2001. Controlling reproduction in domestic mammals. In "La reproduction chez les mammifères et l'homme". "ed INRA, 3ème edition, 792-815.

Cognié J., Baril G., Touze J.L., Petit J.P., 2007. Laparoscopic follow-up of cyclical corpora lutea in ewes. Revue de Médecine Vétérinaire 158 (8-9), 447-451.

Dudouet C., 1997. La production du mouton. Edition France Agricole. 357p

Dudouet C. , 2003. La production du mouton. 2^{éme} édition. Editions France agricole, pp 287. **Fernandez V.E.,** 2003. Technicien en élevage, tome1 et tome2. Cultural, S.A. Madrid. pp 242 and pp 471.

Hassoun P., Bocquier F., 2007. Sheep feed. Alimentation des bovins, ovins et caprins. Edition Quae, INRA, Paris, 121-136.

Jarrige R., 1988. Alimentation des bovins, ovins et coprins. INRA, Paris. pp 476.

Kennedy D., 2002. Off-season breeding of sheep. Data sheet. Avaible from internet:< http://www.omafra.gov.on.ca/french/livestock/sheep/facts/02-064.htm>

Khaldi G., 1984. Seasonal variation in ovarian activity, oestrus behaviour and the duration of post partum anoestrus in Barbarine sheep. Influence of food level and the presence of the male. PhD thesis, Montpellier Academy, pp 168.

Khaldi G., 1989. The Barbary sheep. In "Small ruminants in the Near East". Volume III, FAO, 74, 96-135.

Khaldi G., and Lassoued N., 1991. Interaction nutrition Reproduction chez les ovins en milieu méditerranéen. International Symposium on the application of nuclear and related techniques in animal production and health. In Animal Production and Health, IAEA/FAO, Austria, 1-19 April, Vienna, 379-390.

Khaldi G., 2005. Feeding, production and reproduction of ewes of Barbarine breed in Tunisia. Small Ruminants Training Resource CDROM, ILRI-ICARDA-IRESA. **Khaldi S.,** 2007. Etude des caractères des brebis et de croissance des agneaux de la souche W de la race Barbarine : Résultats de 20 années d'élevage. Doctoral thesis in agronomic sciences, INAT, 120 p.

Lassoued N., and Khaldi G., 1995. Seasonal variation in the sexual activity of Queue Fine de l'Ouest and Noire de Thibar ewes. Sheep farming in arid and semi-arid zones. Cah. Opt. Médit, 6, 27-34.

Lindsay D.R., 1995. The role of management in the control of the estrus cycle. Reproduction and animal breeding: Advances and strategy. Proceedings of the XXX International Symposium of Societa Italiana per il progresso della zootecnica. Eds.Enne.

Ouattara I., 2001. Rapport technique sur Gestion de la reproduction dans un élevage ovin. Institut Agronomique et Vétérinaire HASSAN II. pp 15.

Pelletier J., 1971. Influence of photoperiodism and androgens on the synthesis and release of LH in rams. Th. Doct. Etat ès Sci. nat. Paris N° A 5441.

Rekik M., and Mahouachi M., 1999. Maîtrise de la reproduction et insémination artificielle des ovins E.S.A.K. Tunisie, pp 127.

Theriéz M., 1984. Influence de l'alimentation sur les performances de reproduction des ovins. 9ème Journée de la Recherche Ovine et Caprine. INRA-ITOVIC, p294-326.

Thimonier J., Mauléon P., 1969. Seasonal variations in oestrus behaviour and ovarian and pituitary activity in sheep. Ann. Biol. Anim. Bioch. Biophys. 9, p233-250.

Thimonier J., Cognié Y., Lassoued N., Khaldi G., 2000. L'effet mâle chez les ovins: une technique actuelle de maîtrise de la reproduction. INRA, Prod.Anim. 13(4), p223-231.

Tournadre H., Pellicer M., Bocquier F., 2009, "Maîtriser la reproduction en élevage ovin biologique : influence de facteurs d'élevage sur l'efficacité de l'effet ramer". Innovations agronomiques (2009) 4, 85-90. Avaible from internet : < http://orgprints.org/15453/1/14-Tournadre.pdf>

Zaiem I., Chemli J., Slama H., Tainturier D., 2000. Improving the performance of reproduction by the use of melatonin in counter-seasonal ewes in Tunisia. Revue Méd. Vét, 2000, 151, 6, p517-522.

TABLE OF CONTENTS

Printed by Books on Demand GmbH, Norderstedt / Germany